This Science Fair Notebook Belongs To:

_____

Project Idea Brainstorming
Use These Pages To Jot Down All Your Project Ideas

Project Idea Brainstorming

Project Idea Brainstorming

Can I Make This Project Work?

Use The Following Questions To Determine If Your Project Idea Is A Good One For The Science Fair. The Answer To All Of The Questions Should Be Yes.

	Yes	No
Can I Write A Question For My Topic?	☐	☐
Can It Be Tested?	☐	☐
Do I Have Enough Time To Test It?	☐	☐
Can I Get All The Materials I Need?	☐	☐
Can I Afford The Materials?	☐	☐
Is It Safe?	☐	☐
Do I Fully Understand The Project?	☐	☐
Can It Be Researched?	☐	☐
Is It Original?	☐	☐
Does It Interest Me?	☐	☐

Thoughts

Narrow It Down

Three Testable Questions For My Project

1. _____

_____

_____

_____

2. _____

_____

_____

_____

3. _____

_____

_____

_____

My Project Will Be:

Resources Log

Books/Magazines/Newspapers

Title	Author	Date

Resources Log (Continued)

Websites

URL	Author	Date

Resources Log (Continued)

Personal Interviews

Name	Contact Info.	Date

Research Notes

Research Notes

Research Notes

Research Notes

Research Notes

Research Notes

Research Notes

Research Notes

Research Notes

Research Notes

Research Notes

Research Notes

Research Notes

Research Notes

Research Notes

Research Notes

Research Notes

Research Notes

Research Notes

Research Notes

Planning Form

My Hypothesis

_____

_____

My Variable

_____

_____

My Constant

_____

_____

My Control

_____

_____

How I Will Test My Question

Supplies Needed/Shopping List

Supply	Have	Need To Buy	Can Borrow	Cost

Experimentation Notes

Experimentation Notes

Experimentation Notes

Experimentation Notes

Experimentation Notes

Experimentation Notes

Experimentation Notes

Experimentation Notes

Experimentation Notes

Experimentation Notes

Experimentation Notes

Experimentation Notes

Experimentation Notes

Experimentation Notes

Experimentation Notes

Experimentation Notes

Experimentation Notes

Experimentation Notes

Experimentation Notes

Experimentation Notes

Experimentation Notes

Data Tables

Data Tables

Data Tables

Data Tables

Data Tables

My Project Results

My Results:

My Conclusions:

Project Display Board Sketches

Project Display Board Sketches

Project Display Board Sketches

Project Display Board Sketches

Project Display Board Sketches

Final Report Notes/Rough Draft

Final Report Notes/Rough Draft

Final Report Notes/Rough Draft

Final Report Notes/Rough Draft

Final Report Notes/Rough Draft

Final Report Notes/Rough Draft

Final Report Notes/Rough Draft

Final Report Notes/Rough Draft

Final Report Notes/Rough Draft

Final Report Notes/Rough Draft

Final Report Notes/Rough Draft

Final Report Notes/Rough Draft

Final Report Notes/Rough Draft

Final Report Notes/Rough Draft

Final Report Notes/Rough Draft

Final Report Notes/Rough Draft

Final Report Notes/Rough Draft

Final Report Notes/Rough Draft

Final Report Notes/Rough Draft

Final Report Notes/Rough Draft

Final Report Notes/Rough Draft

Final Report Notes/Rough Draft

Final Report Notes/Rough Draft

Final Report Notes/Rough Draft

Made in the USA
Las Vegas, NV
09 February 2025

17789371R00050